INTERNATIONAL CENTRE FOR MECHANICAL SCIENCES

COURSES AND LECTURES - No. 221

HAYRETTIN KARDESTUNCER

UNIVERSITY OF CONNECTICUT

DISCRETE MECHANICS
A UNIFIED APPROACH

SPRINGER-VERLAG WIEN GMBH

ISBN 978-3-211-81379-9   ISBN 978-3-7091-4350-6 (eBook)

DOI 10.1007/978-3-7091-4350-6

To

Vinicio TURELLO

for his continuous effort in providing a truly academic atmosphere for great masters and students in mechanics of all nations. Such an effort will undoubtedly benefit the scientific world.

PREFACE

Rightfully or not, most researchers in mechanics today have divorced themselves (or tend to) from searching closed-form solutions for their problems. The complicated nature of these problems and the availability of the powerful number-crunching equipment are the main causes of this development. Shortcomings seem to arise, however, if, while observing these rapid changes one disassociates himself from the step stones of the past.

The text introduces very briefly the three most popular approximate methods in their sequential order: calculus of variations, finite difference and finite elements. It then advocates the combination of all three in one physical problem. Each of these three has its own beauty and advantage over the other two yet no one alone is satisfactory for some problems. The initial and the boundary value combinations, particularly, such as path and time dependent problems force us to unify these methods. The text is intended to be a pioneering attempt in this direction.

I am very grateful to Professor W. Olszak, Rector of CISM, for providing me with the opportunity to present this work to the participants of the UNESCO series of lectures at the Center.

H. Kardestuncer

Udine
July 1975

Chapter 1

INTRODUCTION

Ogni azione fatta

dalla natura è fatta

nel più breve modo.

Leonardo da Vinci

Nature always tends to minimize its effort in achieving its goal. Whether it is the maximum area of Queen Dido of Carthage, the minimum resistance of Newton or the shortest time of Bernouilli, most physical problems fall, theoretically at least, within the jurisdiction of variational calculus.

Although the basic concept (minimum, maximum) of variational calculus was known to ancient Egyptians (reallocation of land after the Nile floodings), and Greeks (maximum area or minimum distance), it could not be considered a mathematical discipline until the period of Newton, Leibnitz and the Bernouillis. Much of the formulation of this mathematics was developed by Euler [1] (1707-83) who was a pupil of Jacob Bernouilli.

The earliest accounted problem in this discipline, the brachistochrone problem, was proposed by Johann Bernouilli in 1696 and simultaneously solved by his brother Jacob, Sir Isaac Newton and the French mathematician L'Hopital. Furthermore, Lagrange, a contemporary of Euler, built a new foundation for the subject which later was adopted by Euler to complete his earlier work. The latest major contribution to the subject is due to Weierstrass in about 1875.

The numerical solution of differential equations by the finite differences dates as far back as the era of the Bernouillis and Euler. However, it gained popularity only after the development of high-speed computers. When Courant [2] pointed out the similarities in approximate methods applied to variational and differential problems, the localized Rayleigh-Ritz procedure gave birth to the present finite element method. Similar ideas were used later by Polya [15], Prager [4] and Synge [5].

The following chart illustrates the relationship between the various methods used in mechanics.

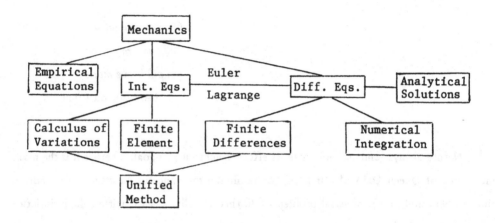

While the development of the approximate methods in physics and in engineering shifted in discrete nature from the variational techniques to finite differences then to finite elements, researchers these days are concentrating their efforts to employ three methods in one physical problem. These efforts will soon result in the most efficient use of *finite calculus* in engineering.

Chapter 2
CALCULUS OF VARIATIONS

The calculus of variations deals with the problem of determining curves or functions $u = u(x)$ such that some functional

$$\pi = \int_{x_1}^{x_2} F(x,u,\dot{u}) \ dx \qquad (2.1)$$

subject to the restriction

$$J = \int_{x_1}^{x_2} G(x,u,\dot{u}) \ dx \qquad (2.2)$$

takes on an external (stationary) value.

The integrand F, in general, may have many dependent and independent variables as well as higher-order derivatives.

$$\pi = \int_{x_1}^{x_2} F\left(x, u_i, u_i^{(1)}, u_i^{(2)}, \ldots, u_i^{(k)}\right) dx \ . \qquad (2.3)$$

It can be shown that the solution of this problem satisfies the following differential equation [4].

$$\sum_{j=0}^{n} (-1)^{n-j} \frac{d^{n-j}}{dx^{n-j}} \left[\frac{\partial F}{\partial u_i^{(n-j)}} \right] = 0 \ , \qquad (2.4)$$

which is known as the Euler-Lagrange differential equation associated with the functional π .

Before going any further, let us consider the simplest problem, i.e., the shortest distance between two points in plane geometry, for which we may employ the variational technique. This problem requires the minimization of the following functional

$$\pi = \int_{x_1}^{x_2} F(x, \dot{u}) \, dx \ ,$$

where

$$F = (1 + u_x^2)^{1/2}$$

is the differential length of the curve.

Equation (2.4) for this problem becomes

$$\frac{\partial F}{\partial u} - \frac{d}{dx} \frac{\partial F}{\partial u_x} = 0 \ .$$

Furthermore,

$$\frac{\partial F}{\partial u} = \frac{\partial}{\partial u} (1 + u_x^2)^{1/2} = 0$$

$$\frac{\partial F}{\partial u_x} = \frac{\partial}{\partial u_x} (1 + u_x^2)^{1/2} = \frac{u_x}{(1 + u_x)^2} \ .$$

Finally

$$- \frac{d}{dx} \left[u_x (1 + u_x^2)^{-1/2} \right] = 0$$

or

$$\frac{u_x}{(1 + u_x^2)^{1/2}} = \text{const.}$$

which implies that

$$u_x = \text{const.}$$

This, of course, represents the straight line between those two points.

In energy problems, for instance, the integral involves the second derivatives in which case Eq. (2.1) is in the following form.

$$\pi = \int_{x_1}^{x_2} F(x,u,u',u'')dx \; . \tag{2.1a}$$

The variation of this integral due to small changes in u can be written as

$$\delta\pi = \int_{x_1}^{x_2} \left(\frac{\partial F}{\partial u} \delta u + \frac{\partial F}{\partial u'} \delta u' + \frac{\partial F}{\partial u''} \delta u'' + \right.$$

$$\left. + \; \text{Higher order terms} \right) dx$$

in which each term can be integrated by parts as

$$\int_{x_1}^{x_2} \frac{\partial F}{\partial u} \delta u \; dx$$

$$\int_{x_1}^{x_2} \frac{\partial F}{\partial u'} \delta u' \; dx = \frac{\partial F}{\partial u'} \delta u \left.\right|_{x_1}^{x_2} - \int_{x_1}^{x_2} \frac{d}{dx} \left(\frac{\partial F}{\partial u'} \right) \delta u \; dx$$

$$\int_{x_1}^{x_2} \frac{\partial F}{\partial u''} \delta u'' \; dx = \frac{\partial F}{\partial u''} \delta u' \left.\right|_{x_1}^{x_2} - \frac{d}{dx} \left(\frac{\partial F}{\partial u''} \right) \delta u \left.\right|_{x_1}^{x_2} +$$

$$+ \int_{x_1}^{x_2} \frac{d^2}{dx^2} \left(\frac{\partial F}{\partial u''} \right) \delta u \; dx \; .$$

The parts which are already integrated are called the *natural boundary conditions*. Either the value of the fuctions or these conditions must be prescribed at the boundaries. In solid mechanics, for instance, the conditions are equivalent to force conditions as opposed to kinematic boundary conditions, i.e., the function or its derivatives, at the boundaries.

$$u, \ u' = 0 \quad \text{fixed end}$$

$$u, \ u'' = 0 \quad \text{hinged end}$$

If we add the corresponding terms

$$\delta\pi = \int_{x_1}^{x_2} \left[\frac{\partial F}{\partial u} - \frac{d}{dx}\left(\frac{\partial F}{\partial u'}\right) + \frac{d^2}{dx^2}\left(\frac{\partial F}{\partial u''}\right) \right] \delta u \ dx$$

$$+ \ \left[\frac{\partial F}{\partial u'} - \frac{d}{dx}\left(\frac{\partial F}{\partial u''}\right) \delta u \right]_{x_1}^{x_2}$$

$$+ \ \left[\left(\frac{\partial F}{\partial u''}\right) \delta u' \right]_{x_1}^{x_2}$$

$$+ \quad \text{Higher order terms}$$

and noting that $\delta\pi$ is arbitrary, then it can only vanish if and only if *every one of its terms vanishes identically*.

$$\frac{\partial F}{\partial u} - \frac{d}{dx}\left(\frac{\partial F}{\partial u'}\right) + \frac{d^2}{dx^2}\left(\frac{\partial F}{\partial u''}\right) = 0 \qquad (2.4a)$$

$$\left[\frac{\partial F}{\partial u'} - \frac{d}{dx}\left(\frac{\partial F}{\partial u''}\right) \right] \delta u \ \Big|_{x_1}^{x_2} = 0 \qquad (2.4b)$$

$$\left[\left(\frac{\partial F}{\partial u''}\right) \delta u' \right]_{x_1}^{x_2} = 0 \ . \qquad (2.4c)$$

Equations (2.4b and c) indicate that the following must be specified at the ends.

$$u \quad \text{or} \quad \frac{d}{dx}\left(\frac{\partial F}{\partial u''}\right) - \frac{\partial F}{\partial u'} = 0 \, ,$$

$$u' \quad \text{or} \quad \frac{\partial F}{\partial u''} = 0 \, .$$

While Eq. (2.4a) is referred to as the Euler-Lagrange differential equation of (2.1a), Eqs. (2.4b) and (2.4c) are the corresponding boundary conditions.

In general, there are three types of boundary conditions:

    a. *Dirichlet type*

    b. *Neumann type*

    c. *Mixed boundary conditions*

While the dependent variable u is prescribed at the boundaries of the first type

$$(u)_s = g_1 \, ,$$

its derivatives are specified in the second type

$$\left(\frac{\partial u}{\partial n}\right)_s = g_2 \, .$$

The combination of the two then yields to the mixed boundary value problem,

$$g_3(u)_s + g_4\left(\frac{\partial u}{\partial n}\right)_s = g_5 \, ,$$

where g_i are the prescribed functions on the boundaries.

The boundary conditions corresponding to the vanishing of the integrated terms in Eqs. (2.4b) and (2.4c) are known as the natural boundary conditions.

16

The variational and the operational forms of Eqs. (2.3) and (2.4) can be written, respectively, as

$$\pi = (L\tilde{u}, \tilde{u}) - 2(f, \tilde{u}) \tag{2.5}$$

and

$$L u = f , \tag{2.6}$$

in which $\tilde{u}$ is an admissible function and L is a positive definite operation. The minimation of π and the inversion of L are equivalent; they produce the same solution u

$$\frac{\partial \pi}{\partial \tilde{u}}\bigg|_{\tilde{u}=u} = 2(L u - f) = 0 .$$

Thus

$$L u = f .$$

Substituting Eq. (2.6) into (2.5) yields

$$\pi = (L\tilde{u}, \tilde{u}) - 2(Lu, \tilde{u})$$

$$= [L (\tilde{u}-u), (\tilde{u}-u)] - L(u,u) ,$$

where "," indicates inner product.

Since L is positive definite, therefore $[L (\tilde{u}-u), (\tilde{u}-u)] \geq 0$ in which the equality holds if and only if $u = \tilde{u}$. In other words, among all possible solutions, it is the exact solution u which makes the functional π a minimum.

The solution procedure in variational calculus starts with introducing an admissible function(s) $a_i \phi_i (x)$ which, in general, satisfies the prescribed boundary conditions. One may then determine the unknown parameters a_i by minimizing the functional

$$\frac{\partial \pi}{\partial a_i} = 0 \quad i = 1,2, \ldots,n \qquad (2.7)$$

or by satisfying the orthogonality condition

$$\int_V (L\hat{u}-f) \, \phi_i \, dv = 0 \qquad i = 1,2,\ldots,n \quad . \qquad (2.8)$$

While Eq. (2.7) is referred to as the Rayleigh-Ritz procedure, Eq. (2.8), which is one of the methods of the weighted residuals, is known as the Galerkin method.

In order to illustrate the solution of the boundary value problems by variational principles, let us consider a simple beam subject to a uniform load p. The total potential energy of this system is

$$\pi = \int_0^L \frac{EI}{2} (u'')^2 \, dx - \int_0^L pu dx' \quad ,$$

where

$$\frac{EI}{2} (u'')^2 - pu = F \quad .$$

According to Eq. (2.4a),

$$- p - 0 + \frac{EI}{2} \frac{d^2}{dx^2} (2u'') = 0 \quad .$$

Thus

$$EI \, u^{IV} - p = 0$$

which is known as the Euler-Lagrange differential equation of the problem. In order to solve the problem in variational form, one must find a suitable set of admissible functions (the functions which satisfy the given boundary conditions). For this problem one may choose it in the form of the Fourier series which certainly satisfy the boundary conditions:

$$u(x) = \sum_{i=1}^{n} a_i \sin \frac{i\pi x}{L} .$$

Consequently:

$$u'(x) = \sum_{i=1}^{n} a_i \left(\frac{i\pi}{L}\right) \cos \frac{i\pi x}{L}$$

$$u''(x) = -\sum_{i=1}^{n} a_i \left(\frac{i\pi}{L}\right)^2 \sin \frac{i\pi x}{L} ,$$

which are the first and second derivatives of the function.

The functional, then, becomes

$$\pi = \frac{1}{2} \int_{0}^{L} EI \sum_{i=1}^{n} \sum_{j=1}^{m} a_i \left(\frac{i\pi}{L}\right)^2 \sin \frac{i\pi x}{L} a_j \left(\frac{j\pi}{L}\right)^2$$

$$\sin \frac{j\pi x}{L} dx - \int_{0}^{L} p \sum_{i=1}^{n} a_i \sin \left(\frac{i\pi x}{L}\right) dx .$$

After the integration, one obtains

$$\pi = \frac{EI(3.14)^4}{4L^3} \left[\sum_{i=1,3,}^{n} i^4 a_i^2 \right] - \frac{2PL}{3.14} \sum_{i=1,3,}^{n} i a_i .$$

Complying with the conditions of minimization

$$\frac{\partial \pi}{\partial a_i} = 0$$

and assuming a numerical value to EI as .25 x 10, the following unknown parameters will result in

$$a_1 = 0.087818$$
$$a_3 = 0.0003614$$
$$a_5 = 0.0000281$$

from which the deflection at the mid-span of the beam can be computed by taking various terms in the series.

$$u_{x = \frac{L}{2}} = a_1 \sin \frac{\pi}{2} = 0.087818$$

or

$$u_{x = \frac{L}{2}} = a_1 \sin \frac{\pi}{2} + a_3 \sin \frac{3\pi}{2} = 0.087457 .$$

Often, in practice, one may not be able to find a suitable set of basis functions which satisfies the prescribed boundary conditions of the entire region. Instead, however, the region of interest can be divided into sub-regions and a set of trial functions, which are independently defined over each sub-region, can be used in the minimization of the functional. This, to which we shall refer as the discrete variational technique, might be considered a transition from the variational methods to the finite element method [9, 11].

Assume, for instance, that the region of interest is divided into N sub-regions and the trial functions are chosen in the form of ortho-normal functions:

$$\tilde{u}_N (x) = \sum_{j=1}^{N} a_j \psi_j (x) . \qquad (2.9)$$

Note that for N = 1 the discrete variational technique becomes identical with the classical variational method. Substituting Eq. (2.9) into (2.5) and observing the ortho-normal properties of $\psi_j (x)$ yields

$$\pi = \sum_{j=1}^{N} [a_j^2 - 2(f, \psi_j) a_j] . \qquad (2.10)$$

The minimization of this functional as

$$\frac{\partial \pi}{\partial a_j} = 0$$

results in

$$a_j = (f, \psi_j) \ ,$$

which concludes that

$$\pi = -\sum_{j=1}^{N} a_j^2 \ . \tag{2.11}$$

This equation indicates that the functional π converges its true value from above without any oscillatory experience as N, the number of sub-regions, increases.

Consider, for instance, the following integral

$$\pi = \int_{o}^{1} [(u')^2 - u^2 - 2xu] \ dx \ , \tag{2.12}$$

which corresponds to the differential equation

$$L \ u = - \ u'' - u = x \ , \tag{2.13}$$

subject to boundary conditions

$$u(o) = u(1) = 0 \ . \tag{2.14}$$

Instead of passing a trial function for $o<u<1$, let the entire region be divided into two equal sub-regions and the trial functions be chosen as

$$u_2 = \begin{cases} a_1 + a_2 x + a_3 x^2 & o \le x \le \frac{1}{2} \\ \\ a_4 + a_5 x + a_6 x^2 & \frac{1}{2} \le x \le 1 \end{cases} .$$

After satisfying the boundary conditions i.e., Eq. (2.14) and the interface compatibilities i.e.,

$$\left. \frac{\partial F}{\partial u'} \right|_{(-1)} = \left. \frac{\partial F}{\partial u'} \right|_{(+)}$$

in which (–) and (+) stand for one or the other side of the interface (in this case it is a point), the trial function, then, becomes

$$u_2 = a_3\, Y_{21} + a_6\, Y_{22} ,$$

where

$$Y_{21} = \begin{cases} -\frac{3}{4} x + x^2 \\ \\ -\frac{1}{4} + \frac{1}{4} x \end{cases} \qquad Y_{22} = \begin{cases} -\frac{x}{4} & o \le x \le \frac{1}{2} \\ \\ \frac{1}{4} - \frac{5}{4} x + x^2 & \frac{1}{2} \le x \le 1 \end{cases} ,$$

in which Y_{21} and Y_{22} are linearly independent. Note that any linear operations on Y_{21} and Y_{22} will neither alter the results nor destroy the linear independency. The trial function u_2 therefore can be expressed as

$$u_2 = a_{21}\, \phi_1 + a_{22}\, \phi_2 \ ,$$

where

$$a_{21} = a_3$$

$$a_{22} = a_6 - a_3$$

$$\phi_1 = -\,(\gamma_{21} + \gamma_{22})$$

$$\phi_2 = -\,4\,\gamma_{22}$$

or explicitly,

$$u_2 = \begin{cases} a_{21}\,(x - x^2) + a_{22}\,(x) & 0 \le x \le \dfrac{1}{2} \\[2ex] a_{21}\,(x - x^2) + a_{22}\,(-1 + 5x - 4x^2) & \dfrac{1}{2} \le x \le 1 \end{cases} .$$

At this moment, for the sake of demonstration, let us assume that the entire region is divided into four sub-regions, then the trial functions in this case would be

$$u_4 = \begin{cases} a_{41}(x - x^2) + a_{42}(x) + a_{43}(x) + a_{44}(x) & 0 \leq x \leq \frac{1}{4} \\[2em] a_{41}(x - x^2) + a_{42}(x) + a_{43}(-1 + 13x - 16x^2) + a_{44}(x) & \frac{1}{4} \leq x \leq \frac{1}{2} \\[2em] a_{41}(x - x^2) + a_{42}(-1 + 5x - 4x^2) + a_{43}(3 - 3x) \\ \qquad\qquad + a_{44}(x) & \frac{1}{2} \leq x \leq \frac{3}{4} \\[2em] a_{41}(x - x^2) + a_{42}(-1 + 5x - 4x^2) + a_{43}(3 - 3x) \\ \qquad\qquad\qquad + a_{44}(-9 + 25x \\ \qquad\qquad\qquad - 16x^2) & \frac{3}{4} \leq x \leq 1 \end{cases}$$

in which one can observe that the trial function u_1 is contained in u_2 and both are contained in u_4. In this manner, one can approximate the actual function with a higher degree of accuracy without using higher order polynomials which is the case in the conventional Rayleigh-Ritz procedure.

Note that as the number of regions increases, the trial functions assume the following pattern

$$u_1 = a_{11}\, \phi_1$$

$$u_2 = a_{21}\, \phi_1 + a_{22}\, \phi_2$$

$$\cdot \ \cdot \ \cdot \ \cdot \ \cdot \ \cdot \ \cdot \ \cdot \ \cdot \ \cdot \ \cdot \ \cdot$$

$$u_N = a_{N1}\, \phi_1 + a_{22}\, \phi_2 + \ \cdot \ \cdot \ \cdot \ \cdot + a_{NN}\, \phi_N$$

$$
\begin{bmatrix} u_1 \\ u_2 \\ \cdot \\ u_N \end{bmatrix} = \begin{bmatrix} a_{11} & & & \\ a_{21} & a_{22} & & \\ \cdot & \cdot & \cdot & \\ a_{N1} & \cdot & \cdot & a_{NN} \end{bmatrix} \begin{bmatrix} \phi_1 \\ \phi_2 \\ \cdot \\ \phi_N \end{bmatrix}
$$

or, shortly,

$$
u_N = a_{Ni} \, \phi_i \qquad i = 1, 2, \ldots, N \quad .
$$

Since ϕ_i are linearly independent, they can be transformed into an energy orthogonal sequence $\theta_1(x)$, $\theta_2(x)$, ... , $\theta_N(x)$ by the process of orthogonalization. Furthermore, they can be normalized into an ortho-normal sequence like $\psi_1(x)$, $\psi_2(x)$, ...$\psi_2(x)$. In other words,

$$
\theta_1(x) = \phi_1(x)
$$

$$
\theta_2(x) = \phi_2(x) - (L\phi_2, \, \psi_1) \, \psi_1
$$

$$
\theta_3(x) = \theta_3(x) - (L\phi_3, \, \psi_1) \, \psi_1 - (L\phi_3, \, \psi_2) \, \psi_2 \quad ,
$$

where

$$
\psi_i(x) = \frac{\theta_i(x)}{(L\theta_i, \, \theta_i)^{\frac{1}{2}}}
$$

The trial functions, then, become as shown in Eq. (2.9).

For the problem posed above the ortho-normal functions are

$$\psi_1(x) = \left(\frac{10}{3}\right)^{\frac{1}{2}} (x - x^2) \qquad 0 \leq x \leq 1$$

$$\psi_2 = \begin{cases} \left(\frac{40}{13}\right)^{\frac{1}{2}} (-x + 2x^2) & 0 \leq x \leq \frac{1}{2} \\ \\ \left(\frac{40}{13}\right)^{\frac{1}{2}} (-1 + 3x - 2x^2) & \frac{1}{2} \leq x \leq 1 \end{cases}$$

One can, therefore, use the trial functions indicated by Eq. (2.9)

$$u_1 = a_1 \, \psi_1(x) \ ,$$

$$u_2 = a_1 \, \psi_1(x) + a_2 \, \psi_2(x) \ .$$

Substituting these into the integral equation and minimizing it in respect to a_j yields

$$a_1 = \frac{1}{12} \left(\frac{10}{3}\right)^{\frac{1}{2}}, \qquad a_2 = \frac{1}{24} \left(\frac{10}{13}\right)^{\frac{1}{2}}$$

which are in agreement with the closed form solution of this problem

$$u(x) = \frac{\sin x}{\sin 1} - x \ .$$

Further details of this procedure are presented in Reference [9].

The solution of this problem by using the Rayleigh-Ritz procedure and Galerkin's method can be found in various texts. For instance, Martin and Carey [27] choose the following function which certainly is admissible for this problem.

$$u = a_i x^i (1-x) \qquad i = 1, 2, \ldots$$

Taking only one term and minimizing π in respect to a_1 yields

$$u_1 = \frac{5}{18} x(1-x) .$$

With two terms, it becomes

$$u_2 = \frac{71}{369} x(1-x) + \frac{7}{41} x^2(1-x) .$$

Galerkin's method refers to the orthogonality of the trial functions, as shown in Eq. (2.8), in which case the solution of this problem reduces to the minimization of

$$\pi = \int_o^1 (-2a_1 + a_1 x - a_1 x^2) (1-x) x dx ,$$

which yields to the same result obtained previously. One certainly can observe that in both the Rayleigh-Ritz and Galerkin methods the degree of accuracy is contingent upon the degree of admissible functions. If, on the other hand, one uses the same trial functions, the results would be identical.

Chapter 3
FINITE DIFFERENCES

Whether the differential equation $Lu = f$ is obtained directly from the physical reasoning or is resulted from the integral equation by the minimization procedure (Euler-Lagrange type), it can be represented in the form of difference equation $L^h U^h = f^h$. It is so hoped that as h approaches zero U^h converges to the true-solution u. The derivatives occurring in the differential equation, then, are approximated in terms of discrete changes rather than the rate of changes. This can be accomplished by using either one-sided differences as in the case of 'backward' ∇ or 'forward' Δ differences

$$\nabla u_i = U_i^h - U_{i-1}^h \quad , \quad \Delta u = U_{i+1}^h - U_i^h \quad , \tag{3.1}$$

or in terms of 'central' differences

$$\delta u_i = U_{i+1/2}^h - U_{i-1/2}^h \quad . \tag{3.2}$$

If one introduces [20] the 'shifting' operator E

$$E \, u_i = U_{i+1} \tag{3.3}$$

and the 'derivative' operator D

$$D \, u_i = \dot{u}_i \quad ,$$

then it follows immediately that

$$\nabla = 1 - E^{-1}, \quad \Delta = E - 1, \quad \delta = E^{1/2} - E^{-1/2} \tag{3.4}$$

and

$$\nabla = 1 - e^{-hD}, \quad \Delta = e^{hD} - 1, \quad \delta = e^{hD/2} - e^{-hD/2} \tag{3.5}$$

Through these, one can establish the difference molecule of the differential equation and introduce it to the interior points of the domain regardless of whether the operator L is linear or non-linear and the boundaries are curved or straight. One must, however, observe that the system is globally stable, that is, U^h depends continuously on f^h as h approaches zero. From here on, one is confronted with the solution of a set of simultaneous equations (often in banded form) and establishing a criteria for error bounds (often difficult).

In order to demonstrate the method and to compare it with the variational procedures, let us treat the simple beam problem presented in Chapter 2. The differential and the integral equations of this problem were, respectively,

$$EI \ u^{IV} - p = 0 \tag{3.6}$$

$$\pi \int_o^L \frac{EI}{2} (u'')^2 \ dx - \int_o^L pudx \tag{3.7}$$

and the boundary conditions are

$$u(o) = u(L) = u''(o) = u''(L) = 0 . \tag{3.8}$$

Since

$$u^{IV} = \frac{1}{h^4} (u_{i-2} - 4u_{i-1} + 6u_i - 4u_{i+1} + u_{i+2}) + \cdots$$

the difference form of the given differential equation with equal subdivision h becomes

$$u_{i-2} - 4u_{i-1} + 6u_i - 4u_{i+1} + u_{i+2} - \frac{P}{EIn^4} = 0 . \qquad (3.9)$$

Let us assume that $n = 6$, that is,

$$h = \frac{1}{6} .$$

Introduction of the boundary conditions, therefore, yields

$$u_1 = u_7 = 0$$

and

$$u_0 = - u_2 , \quad u_8 = - u_6 ,$$

as indicated by Eq. (3.8).

If one now writes the difference molecule Eq. (3.9) at every interior point of the domain and arranges the terms, this will yield the following simultaneous equations:

$$
\begin{bmatrix}
5 & -4 & 1 & 0 & 0 \\
-4 & 6 & -4 & 1 & 0 \\
1 & -4 & 6 & -4 & 1 \\
0 & 1 & -4 & 6 & -4 \\
0 & 0 & 1 & -4 & 5
\end{bmatrix}
\begin{bmatrix}
u_2 \\
u_3 \\
u_4 \\
u_5 \\
u_6
\end{bmatrix}
= \frac{ph^4}{EI}
\begin{bmatrix}
1 \\
1 \\
1 \\
1 \\
1
\end{bmatrix} .
$$

The solution of this set yields

$$u_2 = u_6 = 0.045360165 \, ,$$

$$u_3 = u_5 = 0.077760283 \, ,$$

$$u_4 = u_{x=1/2} = 0.089424327 \, .$$

Note that u_4 represent deflection at mid-span and should be compared to that obtained in the previous chapter.

Let us now solve the problem by using the variational principle over the integral equation Eq. (3.7) instead of the differential equation Eq. (3.6). Substituting the central differences for the derivatives

$$u''_i = \frac{u_{i-1} - 2u_i + u_{i+1}}{h^2} + \ldots$$

Eq. (3.7) takes the following form.

$$\pi = \frac{EI}{2h^4} \int_0^{2h} [u^2_{i-1} - 4u_i \, u_{i-1} + 2u_{i+1} \, u_{i-1} + 4u_i^2$$

$$- 4u_i \, u_{i+1} + u^2_{i+1}] \, dx - \int_0^{2h} p u_i \, dx \, ,$$

where the integration is done from $-h$ to $+h$. The minimization of this integral

$$\frac{\partial \pi}{\partial u_i} = 0$$

yields

$$\frac{EI}{h^3} \begin{bmatrix} 2 & -4 & 2 \\ -4 & 8 & -4 \\ 2 & -4 & 2 \end{bmatrix} \begin{bmatrix} u_{i-1} \\ u_i \\ u_{i+1} \end{bmatrix} = \begin{bmatrix} 0 \\ -2ph \\ 0 \end{bmatrix}.$$

Finally, for $i = 1, 2, \ldots \ldots 7$, one will obtain the following set

$$\frac{EI}{h^3} \begin{bmatrix} 2 & -4 & 2 & & & & \\ -4 & 10 & -8 & 2 & & & \\ 2 & -8 & 12 & -8 & 2 & & \\ & 2 & -8 & 12 & -8 & 2 & \\ & & 2 & -8 & 12 & -8 & 2 \\ & & & 2 & -8 & 10 & -4 \\ & & & & 2 & -4 & 2 \end{bmatrix} \begin{bmatrix} u_1 \\ u_2 \\ u_3 \\ u_4 \\ u_5 \\ u_6 \\ u_7 \end{bmatrix} = 2ph \begin{bmatrix} 1 \\ 1 \\ 1 \\ 1 \\ 1 \\ 1 \\ 1 \end{bmatrix}.$$

The introduction of boundary conditions, i.e., $u_1 = u_7 = 0$, by eliminating the corresponding rows and columns and the solution of the remaining set of equations will be exactly the same as when obtained by solving the differential equation.

Chapter 4

FINITE ELEMENTS

While the analytical methods assume that a continuum is made of infinitesimal elements, the finite element method interprets it as the assembly of finite size elements inter-
connected in a certain fashion. Contrary to the finite difference method which employs the
governing differential equation at discrete points (nodal points), the finite element method
makes use of the governing integral equation at discrete domains (elements). It also differs
from the classical variational methods by attempting to minimize the functional over the
elements instead of over the whole region.

$$\pi = \sum \pi_e = \sum \int_e F(x,y,u,\hat{u}, \ldots) \, de \qquad (4.1)$$

$$\frac{\partial \pi}{\partial U} = \left[\frac{\partial \pi}{\partial U_1} \frac{\partial \pi}{\partial U_2} \cdots \frac{\partial \pi}{\partial U_n} \right]^* = 0 , \qquad (4.2)$$

where U is the nodal value of function u.

A continuum, whether it is one, two, three or even four-dimensional, can always be
divided into sub-regions (elements) provided that the continuity in terms of the function
and its derivatives between the elements is not violated. Although the elements in reality are
connected throughout the interelement boundaries (lines or surfaces), in the finite element
method they are assumed to be connected by discrete points only. The behavior of continuum, therefore, is characterized by the behavior of these elements. Instead of determining the solution for the entire region, piecewise solutions for the individual elements which

are subject to a proper set of imaginary boundary conditions are sought. The solutions within the elements are constructed in terms of boundary values of the function at the nodal points which are, in turn, not known. Often in practice, however, their proper values are determined by utilizing the variational principles. In elasticity problems, for instance, an approximate solution within the element is attained by minimizing certain functionals, i.e., the potential energy, complementary strain energy, etc. Finally, the equilibrium of elements with their adjacent neighbors and the continuity of the localized approximate functions up to certain degrees of derivatives (compatibility conditions) result in the evaluation of the function at the nodal points. In the case of energy functionals, for instance, when the integrand is a quadratic function of u (consequently, that of U)

$$\pi = \int_v (U* \ LU - 2U * f \ )dv$$

Eq. (4.2) then becomes

$$\frac{\partial \pi}{\partial U} = LU + f = 0 . \tag{4.3}$$

This is the equation with which one eventually ends up in the finite element method.

In order to follow the continuity in the development of the finite element method, let us introduce it without making any reference to calculus of variations. The method was originally introduced and developed by engineers for the analysis of structures without really making any reference to variational calculus. We shall therefore start its direct formulation and later point out its close association with variational calculus.

Let us first designate the following.

X Y Z	. . .	*Global coordinate system*
x y z	. . .	*Local coordinate system*
$g_x \ g_y \ g_z$	. . .	*Body forces per unit volume*
$p_x \ p_y \ p_z$	. . .	*Surface forces per unit area*

u v w . . . . *Displacement components in directions*

u(x,y,z) . . . . *Displacement function*

Assume that the body rests in equilbrium under the effect of body and surface forces. Let σ and ε represent the state of stress and strain of the body at rest. If the body is sub-ject to arbitrary virtual displacements Δu (increment in displacements caused by other effects), then the change in the state of strain of the body would be

$$\delta \quad = [\delta\varepsilon_x \ \delta\varepsilon_y \ . \ . \ . \ . \ . \ \delta\varepsilon_{yz}] \ .$$

The internal work done by the actual stresses during the virtual displacements is

$$\delta W_I = \int_V \sigma * \delta\varepsilon \ dv \ . \tag{4.4}$$

The corresponding external work, however, is

$$\delta W_E = \int_V g* \Delta u \ dv + \int_A p* \Delta u \ dA \ , \tag{4.5}$$

where asterisks indicate *"transpose"* of arrays

$$g * = [g_x \ g_y \ g_z] \ ,$$

$$p * = [p_x \ p_y \ p_z] \ .$$

By the conservation of energy, however,

$$\delta W_I = \delta W_E \ . \tag{4.6}$$

Therefore

$$\int_V \sigma* \; \delta\varepsilon \;\; dv \; = \; \int_V g* \; \Delta u \;\; dv \; + \; \int_A p * \Delta u \;\; dA . \qquad (4.7)$$

Although the principle of virtual displacement is independent of material behavior, if the material is linear elastic, then

$$\sigma = D \; (\varepsilon \; - \; \varepsilon_0) , \qquad (4.8)$$

where ε_0 represents the initial strain not associated with actual state of stress. Transposing Eq. (4.7) and substituting Eq. (4.8) into it yields

$$\int_V \delta \; \varepsilon * D \; dv \; = \; \int_V \delta\varepsilon * D \; \varepsilon_0 \; dv \; + \; \int_V \Delta u * g \;\; dv + \int_A \Delta u *p \; dA . \; (4.9)$$

Now assume that U represents a vector associated with the actual displacements of certain points (nodal points) of the body

$$U* \; = \; [\; U_1 \; U_2 \cdot \; \cdot \; \cdot \; \cdot \; \cdot \; U_n \;] . \qquad (4.10)$$

If $u(x, y, z)$ is a continuous single valued function representing the actual displacements of the body, there certainly must exist a relationship between u and U.

$$u \; = \; aU . \qquad (4.11)$$

Since there exists the following strain-displacement relationship (the linear strain tensor)

$$\varepsilon_{ij} \; = \; \frac{1}{2} \; (u_{i,j} + u_{j,i}) \qquad (4.12)$$

by the virtue of Eq. (4.11), there also must exist a relationship between the nodal point displacements and the actual strain pattern of the body.

$$\varepsilon = bU . \tag{4.13}$$

Differentiating Eq. (4.11) in accordance with Eq. (4.12) and substituting it into Eq. (4.9) gives

$$\int_V \Delta U * b * D b U \, dv = \int_V \Delta U * b * \varepsilon_o \, dv + \int_V \Delta U * a * g \, dv$$

$$+ \int_A \Delta U * a * p \, dA . \tag{4.14}$$

Since $\Delta U*$ is arbitrary virtual displacement

$$\int_V b * D b \, dv \, U = \int_V b * \varepsilon_o \, dv + \int_V a * g \, dv$$

$$+ \int_A a * p \, dA , \tag{4.15}$$

from which

$$K U = P_{\varepsilon_o} + P_g + P_p \tag{4.16}$$

or

$$K U = P , \tag{4.17}$$

where

$$K = \int_V b * D b \, dv \qquad \text{\textit{Stiffness matrix of the body}}$$

$$P_{\varepsilon_o} = \int_V b * \varepsilon_o \, dv \qquad \text{\textit{Nodal force vector due to initial straining}}$$

$$P_g = \int_V a * g \, dv \qquad \text{\textit{Nodal force vector due to body forces}}$$

$$P_p = \int_A a * g \, dA \qquad \text{\textit{Nodal force vector due to surface forces}}$$

Equation (4.17) (a more explicit version of Eq. 4.3) which relates the nodal point displacements to the nodal point forces is the main equation in the finite element method. The assembly and solution of this equation is the main theme of this method.

Let us now introduce the method as developed in [24, 28] without again making any reference (explicitly) to variational procedures.

The first theorem of Castigliano states that: If an elastic body is subject to a set of concentrated loads, the first partial derivative of the strain energy in respect to any displacement component of any point is equal to the required force acting at that point in the direction of that displacement.

$$\frac{\partial W}{\partial U_i} = p^i. \qquad (4.18)$$

The strain energy of the body which is equal to the external work can be written as

$$W_I = W_E = \frac{1}{2} p^j U_j \qquad j = 1,2,\ldots\ldots, n. \qquad (4.19)$$

According to Castigliano's theorem, then

$$p^i = \frac{1}{2} \frac{\partial (p^j U_j)}{\partial U_j} = \frac{1}{2} \frac{\partial p^j}{\partial U_i} U_j + \frac{1}{2} \frac{\partial p^j}{\partial U_j} U_j + \frac{1}{2} \frac{\partial U_j}{\partial U_i} p^j ,$$

where the repeated indices indicate summation.

38

Since

$$\frac{\partial U_j}{\partial U_i} = 0, 1 \quad \text{for } i \neq j \text{ and } i = j \text{ respectively},$$

therefore

$$p^i = \frac{1}{2} \frac{\partial p^j}{\partial U_i} U_j + \frac{1}{2} p^i ,$$

from which

$$p^i = \frac{\partial p^j}{\partial U_i} U_j \tag{4.20}$$

or

$$p^i = K^{ij} U_j , \tag{4.21}$$

where

$$K^{ij} = \frac{\partial p^j}{\partial U_i} \tag{4.22}$$

indicating that the elements of the stiffness matrix do in fact represent the variation of forces in respect to displacements, or, more explicitly, the force required to hold point j in place once a displacement is introduced at point i. Evidently, $K^{ij} = 0$ if there are no elements between points i and j. Because of this, the stiffness matrix becomes banded.

If, on the other hand, one refers to Castigliano's first theorem

$$\frac{\partial W}{\partial p^i} = U_i , \tag{4.23}$$

the external work expressed in Eq. (4.19) will yield

$$U_i = D_{ij} \, p^j, \qquad\qquad (4.24)$$

in which D_{ij} is known as the *flexibility matrix* of the system.

Equations (4.21) and (4.24) are dual. In skeletal mechanics they are, respectively, referred to as stiffness (displacement) and flexibility (force) methods of analysis. While the former necessitates the assumption of *displacement functions* within the element, the latter starts with an admissible *stress function*. Depending upon the nature of the problem, one may be preferred over the other. For some problems, however, both schemes may be used simultaneously. This often is referred to as *the mixed method of analysis*.

Although Eq. (4.21) is identical with Eqs. (4.17) and (4.3), it nevertheless attaches more physical meaning to the elements of the stiffness matrix. Such a physical interpretation of the stiffness matrix was not so obvious in Eq. (4.17) because of the undefined array b which is

$$b = \frac{1}{2} \left(\frac{\partial}{\partial x_i} + \frac{\partial}{\partial x_j} \right) a,$$

where a is commonly referred to as the *"displacement function"* or the *"shape function"*. This function plays the most important role in the finite element method that employs stiffness matrices. Its counterpart in the flexibility analysis is the *"stress function"*. In either case, the accuracy of the results depends largely upon their proper choice.

After knowing the basic principles of the finite element method, one may be tempted to say that the finite element solutions may converge to the exact solution by either increasing the degrees of freedom per element or decreasing the element sizes. For many reasons, both may be found impractical. Since, however, the finite element method is a procedure for constructing the solution for the entire domain from the local approximate solutions, the convergence can be attained if the functional itself converges as the size of the element diminishes.

If the main function and its derivatives satisfying the governing differential equation over the entire region is continuous, smooth and single-valued, then its Taylor series expansion (as well as other series) in the vicinity of a point may represent such a function with a fair approximation. The degree of approximation, of course, depends upon the truncation of the series as well as the localization of the expansion. The number of terms in the series is associated with the degree of freedom of elements, and the localization of expansion corresponds to the size of the element. Therefore, the increase of the inter-element continuities and the reduction in the element sizes result in better approximation of the function.

The major portion of reserach in the finite element method goes into the choice of local functions. In fact, wrongly chosen functions may lead to wrong answers in spite of smaller elements and higher degrees of freedom. Although the polynomial expansions due to their simplicity are the most commonly used local functions, as a rule, they do not need to be general. Many authors, in fact, investigated the use of transcendental functions. In general, one should comply with the following criteria while making the choice in continuum mechanics.

1. The local function should be so chosen that the self-straining due to a rigid body motion of the element not be permitted.

2. The local function must represent constant strain (generalized) within the element, otherwise, it will not converge to constant strain (stress) as the element gets smaller.

3. The function must be symmetric so it should not have any preferred directions.

4. The local function need be neither polynomials nor complete. It can, however, be complete of order $r = p - i$ if p represents the order of highest derivative occurring in the energy functional.

5. The local function must be continuous across the inter-element boundaries. The continuity of its derivatives is not necessary provided that the functional itself is not undefined mathematically. If the derivatives as well as the local function itself are continuous

across the inter-element boundaries, such a function is referred to as *"conforming"*, otherwise *"non-conforming"*. Often, in practice, very good results are obtained by non-conforming local functions. Therefore, this requirement should not be so stringent.

Chapter 5

ANALOGY OF THE APPROXIMATE METHODS

So far we have seen that calculus of variations and the finite element method both refer to the integral equations. The finite difference method on the other hand deals primarily with the differential equations. At any stage of the problem however, these three methods follow very analogous paths. In order to illustrate the similarities between them, let us consider an elementary problem such as a simple prismatic beam subject to a distributed transverse load $p(x)$. The integral and the differential equations of this problem are, respectively,

$$\pi = \int \frac{1}{2} \left(EI\ddot{y}^2 - 2\, py \right)\, dx \tag{5.1}$$

and

$$EIy^{IV} = p \, . \tag{5.2}$$

Substituting for y in Eq. (5.1) its equivalence in the form of central differences, the functional (the total potential energy of the system) becomes

$$\pi = \frac{EI}{2} \sum_{i=1}^{n} \frac{(y_{i-1} - 2y_i + y_{i+1})^2}{h^2} - p_i y_i \, ,$$

where the expression for the derivative is the order of h^2.

The stationary value of π, then, indicates that

$$\frac{\partial \pi}{\partial y_i} = 0 = \frac{EI}{h^4} (y_{i-2} - 4 y_{i-1} + 6 y_i - 4 y_{i+1} + y_{i+2} - P_i),$$

(5.3)

which is precisely the finite difference approximation of the differential equation given by Eq. (5.2). One would certainly anticipate this since Eq. (5.2) is the corresponding Euler-Lagrange equation of Eq. (5.1). Because

$$\frac{\partial F}{\partial y} - \frac{d}{dx} \left(\frac{\partial F}{\partial \dot{y}}\right) + \frac{d^2}{dx^2} \left(\frac{\partial F}{\partial \ddot{y}}\right) = 0 ,$$

where

$$F = \frac{EI}{2} \ddot{y}^2 - py ,$$

yields Eq. (5.2).

It can also be shown that the same difference equation [Eq. 5.2] is encountered (implicitly) in the finite element method. Take, for instance, the localized displacement function in the form of a cubic polynomial

$$u = \sum_{i=0}^{3} c_i x^i .$$

(5.4)

After writing the total potential energy of the system

$$\pi = \frac{1}{2} \int \epsilon * D \epsilon \ dx - \int p u \ dx$$

with the stress and strain expressions

$$\epsilon_x = - \frac{\partial^2 v}{\partial x^2} , \quad \sigma_x = D \epsilon_x ,$$

the minimum potential energy principle yields the following stiffness matrix equation for a line element.

$$P_e = \begin{bmatrix} k_{ii} & \vdots & k_{i,i+1} \\ \cdots & + & \cdots \\ k_{i+1,i} & \vdots & k_{i+1,i+1} \end{bmatrix} u_e = EI \begin{bmatrix} \dfrac{12}{L^3} & \dfrac{-6}{L^2} & \vdots & \dfrac{-12}{L^3} & \dfrac{6}{L^2} \\[2mm] & \dfrac{4}{L} & \vdots & \dfrac{-6}{L^2} & \dfrac{2}{L} \\[2mm] \cdots & & + & \cdots & \\[2mm] & & \vdots & \dfrac{12}{L^3} & \dfrac{-6}{L^2} \\[2mm] \text{symm.} & & \vdots & & \dfrac{4}{L} \end{bmatrix} u_e .$$

The assembly of the two adjacent elements at any nodal point i indicates that

$$P_i = k_{i,i-1}\, u_{i-1} + k_{i,i}\, u_i + k_{i,i+1}\, u_{i+1} ,$$

which for the elements of equal length of h becomes

$$\frac{1}{EI}\begin{bmatrix} ph \\ 0 \end{bmatrix} = \begin{bmatrix} \dfrac{-12}{h^3} & \dfrac{-6}{h^2} \\[2mm] \dfrac{6}{h^2} & \dfrac{2}{h} \end{bmatrix}\begin{bmatrix} y_{i-1} \\ \theta_{i-1} \end{bmatrix} + \begin{bmatrix} \dfrac{24}{h^3} & 0 \\[2mm] 0 & \dfrac{4}{h} \end{bmatrix}\begin{bmatrix} y_i \\ \theta_i \end{bmatrix} + \begin{bmatrix} \dfrac{-12}{h^3} & \dfrac{6}{h^2} \\[2mm] \dfrac{-6}{h^2} & \dfrac{2}{h} \end{bmatrix}\begin{bmatrix} y_{i+1} \\ \theta_{i+1} \end{bmatrix}.$$

Substituting the central differences for the slopes

$$\theta_i = \dot{y}_i = \frac{y_{i+1} - y_{i-1}}{2h}$$

and carrying the multiplication, one will obtain

$$\frac{ph}{3} = \frac{EI}{h^4} \, (y_{i-2} - 4y_{i-1} + 6y_i - 4y_{i+1} + y_{i+2}),$$

which possesses the same difference molecule as that of Eq. (5.3).

The problem posed by Eq. (5.1) is solved by the finite element method using six line elements.

The complete stiffness matrix after the introduction of the boundary conditions i.e., $u_1 = u_7 = 0$, takes the following form.

4	-6	2	0											$-ph^2/12$
	24	0	-12	6										ph
		8	-6	2	0									0
			24	0	-12	6								ph
				8	-6	2	0							0
symm.					24	0	-12	6						ph
						8	-6	2	0					0
							24	0	-12	6				ph
								8	-6	2	0			0
									24	0	6			ph
										8	2			0
											4			$ph^2/12$

The solution of this set yields

$$u_2 = u_6 = 0.04428 ,$$

$$u_3 = u_5 = 0.076032 ,$$

$$u_4 = 0.08748 .$$

which represent the displacements at the corresponding points.

These, of course, are in agreement with those obtained before. The discrepancies, however, are expected to increase for large systems since the finite element method requires the solution of equations twice as large as those of the finite difference scheme.

Let us now consider the heat flow problem where $u(x, y, z, t)$ represent the temperature distribution in the domain. In the absence of sources the corresponding continuity equation is

$$\nabla^2 u = \frac{1}{\alpha} \frac{\partial u}{\partial t} , \qquad (5.5)$$

where α is referred to as the thermal diffusivity of the medium.

The variational form of this equation for a one-dimensional case with $\alpha = 1$ is

$$\pi = \int u_x^2 \, dx + \frac{d}{dt} \int \frac{1}{2} u^2 \, dx . \qquad (5.6)$$

Note that while the steady state heat flow is represented by a, an elliptic partial differential equation, and is a pure boundary-value problem, the unsteady case, Eq. (5.5), is a parabolic-differential equation with boundary and initial values.

The difference molecule of Eq. (5.5) can be written as

$$\frac{u_{i-1}^n - 2u_i^n + u_{i+1}^n}{h^2} - \frac{u_i^{n+1} - u_i^n}{\Delta t} = 0 , \qquad (5.7)$$

in which n and i identify respectively the particular time and location (space).

The repetition of this molecule at every interior point of the domain and the introduction of the boundary conditions yield a set of simulataneous equations which eventaully results in the temperature gradient of the domain.

Instead, however, let us now deal with the integral equation and observe the similarities between the two. Using first order forward differences for the derivatives,

$$\hat{u}_i = \frac{u_{i+1}^n - u_i^n}{h}$$

Equation (5.6) becomes

$$\pi = \int_o^h \frac{u_{i+1}^{n2} - 2u_{i+1}^n u_i^n + u_i^{n2}}{h^2} \, dx + \frac{d}{dt} \int_o^h \frac{1}{2} u_i^{n2} \, dx \; .$$

Integrating and minimizing it, one will obtain

$$\frac{\partial \pi}{\partial u_i} = \frac{2(-u_{i+1}^n + u_i^n)}{h^2} h + \frac{d}{dt} u_i^n \; h = 0 \; .$$

Substituting

$$\frac{d}{dt} u_i^n = \frac{u_i^{n+1} - u_i^n}{\Delta t} \; ,$$

the previous equation becomes

$$\frac{\partial \pi}{\partial u_i} = \frac{2(u_i^n - u_{i+1}^n)}{h^2} h + \frac{u_i^{n+1} - u_i^n}{\Delta t} h = 0$$

or

$$\frac{-2u_i^n + 2u_{i+1}^n}{h^2} - \frac{u_i^{n+1} - u_i^n}{\Delta t} = 0$$

representing the molecule to be written at the interior points of the domain. Again n and i identify the particular time and location.

For i = 1, 2, ... n, this equation takes the following form

$$\frac{1}{h^2} \begin{bmatrix} -1 & 1 & & & & \\ 1 & -2 & 1 & & & \\ & 1 & -2 & 1 & & \\ & & 1 & -2 & 1 & \\ & & & & & 1 \\ & & & & 1 & 1 \end{bmatrix} u^n - \frac{1}{\Delta t} [u^{n+1} - u^n] = 0 ,$$

which is the same as Eq. (5.7) written at the interior points.

Note that in either case (operational or variational) u^{n+1} can be explicitly determined from u^n (initial condition). This is often referred to as the explicit scheme and is subject to stability condition

$$\frac{\Delta t}{h^2} \leq \frac{1}{2} ,$$

since

$$u_i^{n+1} = \frac{\Delta t}{h^2} \{ u_{i-1}^n - \left(2 - \frac{h^2}{\Delta t} \right) u_i^n + u_{i+1}^n \} .$$

This condition requires use of a very small time increment. One can, on the other hand, write Eq. (5.7) at the time (n + 1)

$$\frac{u_{i-1}^{n+1} - 2u_i^{n+1} + u_{i+1}^{n+1}}{h^2} - \frac{u_i^{n+1} - u_i^n}{\Delta t} = 0 ,$$

from which u_i^{n+1} can only be determined from u_i^n by solving a set of simultaneous equations at each time step

$$\frac{\Delta t}{h^2} \left[1 - \left(2 + \frac{h^2}{\Delta t} \right) \; 1 \right] \begin{bmatrix} u_{i-1} \\ u_i \\ u_{i+1} \end{bmatrix}^{n+1} = u_i^n .$$

This equation is a diagonally dominant system and can be solved very efficiently by any one of the direct methods.

Although explicit schemes are often used for elliptic and parabolic equations, for hyperbolic systems such as the wave equation

$$u_{tt} - c^2 \nabla^2 u = f(x,y,z,t)$$

the implicit scheme can be used with much success. Richtmeyer and Morton [19] discuss this problem at length. The price paid for the implicit scheme, of course, is the solution of a large system of equations at each time step which may compensate for the danger of instability of the explicit scheme. A thorough study of implicit versus explicit methods can be found in Part III of reference [33]. A comparison of the finite element method and finite differences method is given by Pian [10]. Furthermore, Kay and Krieg [33] have compared and pointed out the similarities of the two methods for one-dimensional wave motion.

Chapter 6

UNIFICATION OF APPROXIMATE METHODS

It seems that the approximate analysis of physical problems started with the calculus of variations, was raised with the finite differences and matured with the finite element method, depending upon the nature of a problem; however, each has its own merit and sometimes neither one alone is sufficient. Preference for one over the other two often becomes evident at the start (during the formulation of the problems). For some problems, for instance, the functional is extremely difficult to formulate. Going from the functional form to the differential form can always be achieved by the Euler-Lagrange minimization procedure; the reverse, however, is often not possible. Yet for some problems the direct formulation of the differential equation is easier. Most linear solid mechanics problems, for instance, can be handled by the direct stiffness or flexibility approach without referring to the minimization techniques explicitly.

In general, the physical problems can be categorized as

      I. *Linear, static problems*

     II. *Path dependent problems*

   III. *Time dependent problems*

   IV. *Path-Time dependent problems*

For the problems in the first category which are strictly boundary value problems, the choice between the three methods is controlled by factors such as the boundary conditions, convergence and stability, and machine time.

For classical problems, the text book type such as the one just mentioned in the previous article, a variational method may be found far superior to that of finite differences and finite elements provided, of course, that the boundary conditions are not irregular either physically or geometrically. Otherwise, the finite difference method is advantageous over the finite element method since it involves a smaller number of equations to solve, and is disadvantageous because it requires the modification of the molecule near the boundaries and for unevenly spaced mesh points. Actually, this is one of the main reasons that the finite element method gained popularity.

The problems in the last three categories can be interpreted as the combination of boundary and initial value problems. The path dependent problems include: geometric, material and loading nonlinearities, creep, contact and friction problems, crack propagation, etc. All transient field problems such as unsteady vibration, diffusion, heat conduction, seepage and irrotational fluid flow, etc. fall into the third category. The last category is basically the combination of problems in the second and third categories such as elastoplastic and non-linear vibrations, etc. Direct variational methods alone are seldom employed for these problems because of the required discouraging computational effort. They can, however, be implicitly implemented in the finite element method. For instance, Bushnell [30] has introduced with great success the finite difference energy method in the form of discrete variational technique to non-linear shell analysis.

Most physical problems lie in three dimensional real space. All, with the exception of those in the first category, are subject to some history which can be observed in the fourth dimension (path or time axes). One can, then, either discretize the system with time as one of the dimensions of the elements as done by Oden in [25] or use only real elements with incremental or marching technique in the fourth dimension. The latter, which employs the finite element method and finite differences method, respectively, in the boundary and the initial value portions of the same problem, seems very promising. In doing this, the finite element technique in the real domain is carried on to the time domain by the finite difference scheme.

The finite element in time has been investigated by many [23, 25, 26]. The similarities, rather than the unification of the finite element method and finite differences method, are presented by Croll and Walker [31] for plane stress problems. They point out that the molecules of the localized Ritz procedure with partially conforming functions and the conventional finite difference methods are the same. In fact, such similarities were observed earlier in [3, 15, 16]. The unification of the two for problems in category I is not as signifi-cant as for time and path dependent problems. Bushnell [33] has shown that certain finite-difference models are equivalent to constant strain elements with normal displacement and slope discontinuities. He then attempted to give physical interpretation in the form of real elements to finite difference models. Such an interpretation, which may not be essential to mathematicians yet, will be most welcomed in the engineering community. This is one of the reasons that finite difference methods were overshadowed by the finite element method.

As we mentioned earlier, whether the problem is path dependent or time dependent, it can be interpreted as an initial value problem. Consider for example the following non-linear set of equations resulted from the finite element idealization of a problem.

$$K (u) \ u \ = \ p \ (u).$$ (6.1)

Since K and p, the stiffness matrix and load vector, respectively, are functions of u they are not known in advance at different stages of the problem. Therefore, for an arbitrary dis-placement vector u_i this equation will not be satisfied. Let it be represented as

$$F_i = K_i \ u_i - P_i \neq 0 \ ,$$ (6.2)

where the subscript denotes the level of interation.

The Taylor series expansion of F

$$F_{i+1} = F_i + \frac{\partial F}{\partial u_i} (u_{i+1} - u_i) + \frac{\partial^2 F}{2\partial u_i^2}(u_{i+1} - u_i)^2 + \ . \ . \ .$$

indicates that

$$u_{i+1} - u_i = \left[\frac{\partial F}{\partial u_i}\right]^{-1} (F_{i+1} - F_i) + \text{higher order terms} \ ,$$

where the first term on the right-hand side of this equation is referred to as the *Jacobian* which always possesses an inverse.

The right-hand side of this equation can be equated to a multiple of the original equation

$$u_{i+1} - u_i = C_i (k_i u_i - p_i) , \qquad (6.3)$$

where C_i is a relaxation parameter matrix which can be used as analogous to time parameter t, then, Eq. (6.3) represents the following differential equation.

$$\frac{\partial u}{\partial t} + K u - p = 0 . \qquad (6.4)$$

It is interesting to note that path dependent problems can be treated in a manner similar to that of time dependent problems. In solving time dependent problems, as pointed out by Wright and Baron [33], the most difficult decision is whether to use an *explicit* or an *implicit* scheme. While the former is subject to stability consideration, the latter requires the solution of a set of simultaneous equations at each step.

There are numerous procedures for the solution of first order differential equations. Higher order differential equations can always be reduced to a set of first order simultaneous equations. Consider, for instance, the following equation which represents forced vibration of an elastic system with damping.

$$M\ddot{U} + C\dot{U} + K U = P (t) . \qquad (6.5)$$

After bringing this into two first order differential equations

$$\dot{V} = -M^{-1} C V - M^{-1} K U + M^{-1} P$$
$$\dot{U} = V$$

and introducing the fourth order Runge-Kutta's coefficient in the form of

$$\alpha_1 = -M^{-1} C_j V_i - M^{-1} K_j U_i + M^{-1} P_i$$

54

$$\beta_1 = V_i$$

$$\alpha_2 = -M^{-1}C_jV_j - M^{-1}K_j(U_i + 1/2\,V_i\Delta t) + M^{-1}P_i + \frac{\Delta t}{2}$$

$$\beta_2 = V_i + 1/2\,\alpha_1\,\Delta t$$

etc.

one will obtain

$$V_{n+1} = V_n + \frac{\Delta t}{6}(\alpha_1 + 2\alpha_2 + 2\alpha_3 + \alpha_4)\,,$$

$$U_{n+1} = U_n + \frac{\Delta t}{6}(\beta_1 + 2\beta_2 + 2\beta_3 + \beta_4)\,.$$

When material and damping non-linearities exist, the corresponding stiffness and damping matrices need to be evaluated at each time step. The Newton-Raphson method which refers only to the first derivative of the load-deflection curve is the most common procedure for this purpose. The derivation of non-linear stiffness matrices and the use of the Newton-Raphson method along with other initial value procedures can be found in [35] and [32, 37] respectively.

Note that the stiffness matrix K in Eq. (6.2) can be assembled either using finite element or finite differences techniques. Bushnell [33] even recommends using both of these techniques in different regions of the same problem. In other words, the two techniques can alternately be employed during the formulation and the solution of the same problem.

For example, when the finite difference and finite element method (without modification) are combined in a physical domain, the corresponding matrix takes the following form

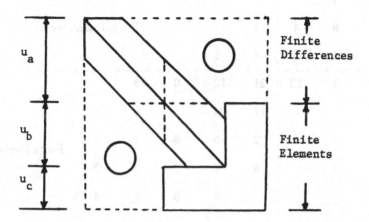

The first portion of this array is formed by finite differences and the second and third by the finite element method. Notice that the bordered form of the second half of this array is due to the fact that the compatibilities in terms of the derivatives of the function are satisfied at the nodal points. If one uses the finite difference technique throughout this problem, the order of this array prior to the introduction of the boundary conditions would have been $(u_a + u_b)$, whereas it is $n u_a + u_b + u_c$ in the finite element method where n represents the degree of freedom (nodal point unknowns) of the system. In addition to this, the band width would increase by one (in sub-matrices terms).

Notice that while the regions u_a and u_b deal with the function itself, the third region involves its derivatives. Notice further that the symmetry of the entire array is disturbed at the interfaces since the finite element molecule is modified to tie to the finite difference molecule at these locations. Such a modification is done by replacing the derivatives in terms of the differences. Farther away from the interfaces (in both directions), however, the symmetry is preserved. Since the nodal points on the interfaces are few, a symmetric solution technique can successfully be employed for the major portion of this array.

For the purpose of illustration, the simple beam problem posed in Chapter 5 is re-solved by using Equations (5.1) and (5.2) with the interface at point 5. The corresponding array takes the following form after the boundary conditions are introduced.

5	- 4	1						
- 4	6	- 4	1					Finite Difference
1	- 4	6	- 4	1				
	3	-12	21	-12	0	6		
		0	-12	24	- 6	0	6	
		4	2	- 6	8	2	0	Finite Element
			6	0	2	8	2	
			6	0	2	4		

After evaluating the vector of constants (force vector in the case of solid mechanics problem), the solution of the set of equations formed by the above array should yield the unknown nodal values of the function. The reader is advised to compare this array with its counterparts presented by the finite difference and finite element methods.

REFERENCES

ON CALCULUS OF VARIATION:

1 Euler, L., *Methods Inveniendi Lineas Curvas Maximi Minimive*, Proprietate Gaudentes, Bousquet, Lausanne and Geneva, 1744.

2 Courant, R., and Hilbert, D., *Methods of Mathematical Physics*, Vol. 1, 1st English ed., 4th Printing, 1963. Wiley (Interscience), New Yorᴸ 1953.

3 Tonelli, L., *Fondamenti di Calcolo delle Variazioni*, Vol. I, II, Bologna, 1921-23.

4 Prager, W. and Synge, J. L., Approximation in elasticity based on the concept of function space, Quarterly App. Math., Vol. 5, 1947.

5 Synge, J. L., *The Hypercircle in Mathematical Physics*, Cambridge University Press, London and New York, 1957.

6 Lanczos, C., *The Variational Principles of Mechanics*, University of Toronto, 1949.

7 Schecter, R. S., *The Variational Method in Engineering*, McGraw-Hill Book Company, 1967.

8 Washizu, K., *Variational Methods in Elasticity and Plasticity*, Pergamon Press, 1968.

9 Hsu, J. M., A discrete variational method for certain boundary value problems, PhD Thesis, University of Connecticut, 1969.

10 Pian, T. H. H., Variational formulations of numerical methods in solid continuum, SMD Symp. Comput. Aided Eng. Univ. of Waterloo, Waterloo, Ontario, Canada, May 1971.

11 Kardestuncer, H., and Hsu, J. M., A discrete variational method for certain boundary value problems in continuous media, V IKM, Weimar, 1969.

12 Weinstock, R., *Calculus of Variations*, McGraw-Hill Book Company, New York, 1975.

13 Kantorovich, L. V., and Krylov, V. I., *Approximate Methods of Higher Analysis*, Interscience Publishers Inc., New York, 1958.

ON FINITE DIFFERENCES:

14 Collatz, L., Konvergenz des differenzverfahrens bei eigenwertproblemen partieller differentialgleichungen, deutsche Math. 3, 200-212, 1938.

15 Polya, G., Sur une interpretation de la methode des differences finies qui puet fournir des bornes superieures ou inferieures, C. R. Acad. Sci. Paris 235, 995-997, 1952.

16 Weinberger, H. F., Upper and lower bounds for eigenvalues by finite difference methods, Comm. Pure Appl. Math. 9, 613-623, 1956.

17 Forsythe, G. E., and Wasow, W. R., *Finite Difference Methods for Partial Differential Equations*, Wiley, New York, 1960.

18 Salvadori, M. G., and Baron, M. L., *Numerical Methods in Engineering*, 2nd ed. Prentice-Hall, Englewood Cliffs, New Jersey, 1961.

19 Richtmyer, R. D., and Morton, K. W., *Difference Methods for Initial Value Problems*, 2nd ed., Wiley (Interscience), New York, 1967.

20 Hildebrand, F. B., *Finite Difference Equations and Simulations*, Prentice-Hall, 1968.

ON FINITE ELEMENTS:

21 Argyris, J. H., and Kelsey, S., Energy theorems and structural analysis, Aircraft Eng. 26-27, Oct. 1954 - May 1955.

22 Turner, M. J., Clough, R. W., Martin, H. C., and Topp, L. J., Stiffness and deflection analysis of complex structures, J. Aeron. Sci., Vol. 23, No. 9, 805-824, Sept. 1956.

23 Zienkiewicz, O. C., and Cheung, Y. K., *The Finite Element Method*, McGraw-Hill, New York, 1967.

24 Kardestuncer, H., *Finite Element Method Via Tensors*, Springer Verlag, CISM, Udine, 1972.

25 Oden, J. T., *Finite Elements of Nonlinear Continua*, McGraw-Hill, New York, 1972.

26 Strang, G., and Fix, G., *An Analysis of the Finite Element Method*, Prentice-Hall, 1974.

27 Martin, H. C. and Carey, G. H., *Introduction to Finite Element Analysis*, McGraw-Hill, 1973.

28 Kardestuncer, H., *Elementary Matrix Analysis of Structures*, McGraw-Hill, 1974.

ON UNIFICATION OF THE METHODS:

29 Wempner, G. A., Finite differences via finite elements, panel discussion of finite elements versus finite differences, Conf. Comput. Oriented-Anal. Shell Struct., Lockheed Palo Alto Res. Lab., Palo Alto, California, Aug. 1970.

30 Bushnell, D., and Almroth, B. O., Finite difference energy method for nonlinear shell analysis, paper presented at Lockheed Missiles and Space Co. Symp., Aug. 1970.

31 Croll, J. G. A., and Walker, A. C., The finite difference and localized Ritz methods, Int. J. Numer. Methods Eng. 3, 155-160, 1971.

32 Stricklin, J. A., Haisler, W. E., and Von Riesemann, W. A., Formulation, computation, and solution procedures for material and/or geometric nonlinear structural analysis by the finite element method, Sandia Laboratories, SC-CR-72-3102, July 1972.

33 Fenves, Perrone, Robinson and Schnorrich (eds.), *Numerical and Computer Methods in Structural Mechanics*, Academic Press, New York, 1973.

34 Hartung, R. F. (ed.), Numerical solutions of nonlinear structural problems, ASME. AMD-Vol. 6, 1973.

35 Rajasekaran, S., Murray, D. W., Incremental finite element matrices, ASCE, Vol.99, ST12, Dec. 1973.

36 Felippa, C. A. (Discussion on Reference 35), ASCE, ST 12, Dec. 1974.

37 Kao, R., A comparison of Newton-Raphson methods and incremental procedures for geometrically nonlinear analysis, Computers and Structures, Vol. 4, Pergamon Press, 1974.

CONTENTS

Printed in the United States
By Bookmasters